**TRURO SCHOOL
PHYSICS DEPT.**

LONGMAN PHYSICS TOPICS *General Editor:* John L. Lewis

# CRYSTALS

Sister M. M. Hurst
(*Sister St Joan of Arc*)

*Formerly Team Leader and Area Coordinator,
Nuffield O-Level Physics Project*

Illustrated by T. H. McArthur

GW00630771

LONGMAN

LONGMAN GROUP LIMITED
London
*Associated branches, companies and representatives throughout the world*

© *Longman Group Ltd (formerly Longmans, Green and Co Ltd) 1969*

*First published 1969*
*Fifth impression 1976*
*ISBN 0 582 32176 X*
*Printed in Hong Kong by*
*Sheck Wah Tong Printing Press*

**ACKNOWLEDGEMENTS**

The author and publisher are grateful for permission to use the following photographs:

Colour plates 2, 4, 6a, 6b, 7, 8 and 9: Geological Survey and Museum, London (Crown copyright reserved). Colour plate 3 by courtesy of Budd Instruments.

Photographs on pages 5, 18, 19 and 20: Geological Survey and Museum, London (Crown copyright reserved). Page 8: (*right*) Imperial Chemical Industries Ltd; (*left*) Cerebos Foods Ltd. Page 9: British Travel Association. Pages 11 and 13: Science Museum, London. Page 15: Novosti Press Agency. Pages 16 and 17: Industrial Diamond Information Bureau. Page 17 (*bottom left*): British Museum (Natural History). Photographs on pages 23, 25, 29 and 48 are by Derek Balmer. Photographs on pages 31 and 33 are from Martin, *Thirteen Steps to the Atom* (Harrap). Page 39: Meteorological Office. Page 39 (*bottom*): from *Physics and Chemistry of Water* by courtesy of Unilever Films. Pages 40 and 41: by courtesy of Mr L. Phoenix. Pages 42 and 43: by courtesy of Mr H. E. C. Powers (retired Tate & Lyle Ltd). Page 45: National Physical Laboratory (Crown copyright reserved). Page 50: (*top*) Mullard Ltd; (*bottom*) Philips Electrical Ltd. Page 51: United States Information Service. Page 52: Sean Connery as James Bond in 'Goldfinger' by courtesy of United Artists Corporation Ltd. Page 54: British Broadcasting Corporation. Page 56: Bell Telephone Laboratories. Page 57: G. & E. Bradley Ltd.

The diagrams at the foot of page 15 are taken by permission of the publishers from Marrison, *Crystals, Diamonds and Transistors* (Penguin); the diagrams on page 18 are from Dana, *Minerals and How to Study Them* (Wiley) and those on pages 46 and 47 are from Mann, *Revolution in Electricity*, copyright © 1962 by Martin Mann (all rights reserved), reproduced by permission of the Viking Press Inc.

We should also like to thank Mr John Burls of the Industrial Diamond Information Bureau for his help and advice.

**NOTE
TO THE
TEACHER**

This book is one in a series of physics background books intended primarily for use with the Nuffield O-Level Physics Project. The team of writers who have contributed to the series were all associated with that project. It was always intended that the Nuffield teachers' materials should be accompanied by background books for pupils to read, and a number of such books is being produced under the Foundation's auspices. This series is intended as a supplement to the Nuffield pupils' materials: not books giving the answers to all the investigations pupils will be doing in the laboratory, certainly not textbooks in the conventional sense, but books that are easy to read and copiously illustrated, and which show how the principles studied in school are applied in the outside world.

The books are such that they can be used with a conventional as well as a modern physics programme. Whatever course pupils are following, they often need straightforward books to help clarify their knowledge, and sometimes to help them catch up on any topic they have missed in their school course. It is hoped that this series will meet that need.

This background series will provide suitable material for reading in homework. This volume is divided into sections, and the teacher may feel that one section at a time is suitable for each homework session for which he wishes to use the book.

This book is written as a background book for the work on crystals in Year 1. The ideas introduced range well beyond those touched on in class, and it is hoped that the questions raised will impel the pupils to further experiment, to consult encyclopedias and other books and to torment their elders! An attempt (preposterous, some would say) is made to stress the nature of scientific activity, to bring the young scientists to the very frontiers of research where problems are still crying for solution, and to encourage and inspire them to cultivate the honest and enquiring mind which is at the root of all intellectual endeavour.

# CONTENTS

# CRYSTALS IN ROCKS

These lovely crystals of white quartz were found in the old hard rocks of Cornwall.

And these blue cubes of fluorite were quarried in the caves of the Peak District.

You may see vases and ornaments (like the one in the colour plate) made of 'Blue John', which is an old name for fluorite.

*Fluorite crystals.*

These almost transparent crystals of calcite come from Cumberland. Why are they shaped like this? How did they come to be in the rocks of the earth?

Quartz crystals are made of the same material as common sand. Chemists call it silicon dioxide. But sand grains on the sea shore are not this shape. Why do they look so different?

5

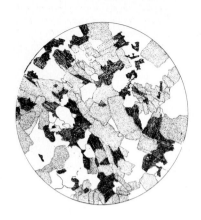

In fact, the grains of sand were once recognizable crystals of quartz in solid rock. Wind and rain, heat and cold over millions of years have broken up the rock; the crystals rubbing against each other in beds of streams and under the sea have been worn down to the tiny grains of sand. They no longer look like crystals because their sharp edges and faces have been rubbed off.

*This is a photograph of a thin slice cut from a piece of Cornish granite.*

You may wonder how anyone can cut a thin slice of something as hard as granite. It is done by a carborundum wheel, which cuts through the rock rather like a steel wheel cutting through bacon on a slicing machine.

You can see that the rock is made up of three minerals: biotite, which looks brown in the section and is dark and shiny in the solid rock; a white mineral called orthoclase feldspar, which looks grey; and clear crystals of quartz.

Some granites are among the oldest rocks of the earth's surface. They are believed to have come from the roots of old mountain chains where the rock was so hot that it was molten. It flowed into cracks in the rock like a liquid, and then as it cooled it turned back slowly to a solid. Rocks formed in this way are called 'igneous', meaning 'fire-formed'. When a volcano erupts molten rock runs out, but it cools quickly and sometimes forms the rough, light stone known as pumice. It is really volcanic lava.

## CRYSTALS FROM MOLTEN ROCK

When the molten rock flows into cracks its temperature is about 1200°C. As it cools various minerals form. In granite the feldspars were the first to solidify so, having plenty of space, they formed large, well-shaped crystals. Mica was the next to solidify as the rock gradually cooled, and then, last of all, the quartz. This meant that the quartz had to fill in the spaces between the other minerals, so it could not form well-shaped crystals. The quartz crystals on page 5 were found in a hole in rock where there was room for them to develop. Notice how the free ends of the crystals are the same shape, although they differ in size.

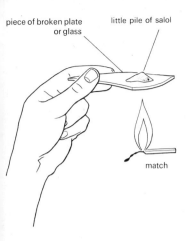

piece of broken plate or glass

little pile of salol

match

*You can warm the glass in very hot water, or over a match.*

## Something to do

If you would like to see crystals forming 'out of the melt' as the granite crystals were formed, use salol, a substance which you can buy from the chemist. Put a spoonful into a small glass bottle with a screw top and heat the bottle in a pan of water. The salol melts at 43°C, which is not much above body temperature.

The water will have to be hotter than that, if you are heating the salol in a bottle. Can you think why?

When it is melted, lift the bottle out of the water and watch it cool. Through the glass sides of the bottle you will be able to see the crystals forming.

Another way to do this is to melt a little on a piece of glass or porcelain.

Blow out the match as soon as the salol melts, and watch. As the molten material cools the crystals grow. Use a hand lens to look at them. They form first at the edges of the little pool of liquid and here they seem to grow to a particular shape. But the ones which form later are restricted by each other's growth, and are irregular shapes.

## Crystals from solutions

As the hot molten rock, 'magma' as it is called, continues to cool, more and more minerals crystallise from it. By the time it has cooled to about 500°C it contains water in which some minerals are still dissolved.

### Something to find out

Water boils at 100°C, so at 500°C you would expect it to have turned to steam. But it hasn't. Why not? Do you know why food cooks more quickly in a pressure cooker? The same fact will provide the answer to both questions.

As the hot solution cools still more and finds its way into the surface rocks, the water evaporates and the minerals still dissolved in it also crystallise out. They form large crystals if the process is slow, but only small ones if it is relatively quick.

## SALT SOLUTIONS AND CRYSTALS

Common kitchen salt was formed like this.

About 200 million years ago there were desert conditions on parts of the earth, a drought so bad that in places even the sea water dried up, and great layers of sea salts were left behind. Later they became thickly covered by wind-blown dust which protected them when huge, slow earth movements caused the land to sink beneath the sea. Layers of older, worn-down rock were brought down by rivers and piled on top of the salt, and these sediments became compressed into solid rock as more and more layers were laid down. Later still, the land rose again, and today we find these deposits of salt under the ground in Cheshire and other places. The rock in which they are found is called the New Red Sandstone. Salt is also found in the Old Red Sandstone, which is about 400 million years old. Sometimes it is mined; sometimes water is pumped down to dissolve it, and then the solution can be pumped up to the surface again.

### Something to think about

How do you think the mining companies get back the solid salt from this solution?

Look at some grains of common salt under a hand lens or a microscope. What shape are they? In class you will have seen a calcite crystal cleaved. Rock salt can be cleaved too, but the cleaved faces, unlike those of calcite, are at right angles to each other: the salt breaks into little cubes.

Below right: *A salt mine.*

Below: *Crystals of common salt.*

Calcite crystals too, like those on page 5, have been formed by crystallisation out of a solution.

The chemical name for calcite is calcium carbonate. Chalk, limestone, marble and most shells are also made of calcite, though the crystals are too small and closely packed to be seen.

Calcium carbonate is not very soluble in water, so pure water running over limestone does not make a solution of it. But when rain falls through the air some carbon dioxide gas dissolves in it, making it into a weak acid. This weak acid reacts with the limestone.

So rainwater running down the hillsides as streams and rivers gradually eats away limestone, especially at the joints between blocks, making great caves. Have you ever been inside Cheddar Caves or Wookey Hole?

*Cox's Cave, Cheddar, showing stalactites, formed where water drips from the roof, and stalagmites, growing up where drops fall to the ground.*

Just as you can get back salt that has been dissolved in water by evaporating the water, so do you get back the original calcium carbonate (and also the carbon dioxide gas) when the water evaporates as it drips from the roof and sides of caves. But it evaporates very slowly, so the calcium carbonate has time to grow into crystals, which you can just see if a stalactite is broken. The calcite crystals in the photograph have had plenty of time as well as plenty of space to become so well-formed.

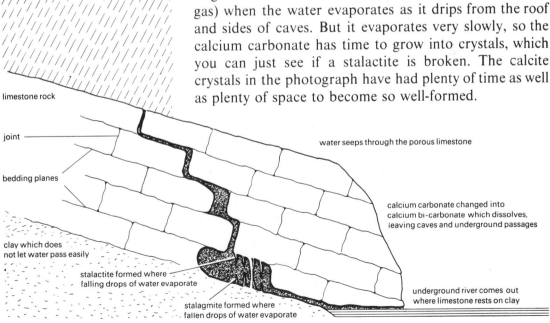

rain water dissolves carbon dioxide from the air

limestone rock

joint

bedding planes

water seeps through the porous limestone

calcium carbonate changed into calcium bi-carbonate which dissolves, leaving caves and underground passages

clay which does not let water pass easily

stalactite formed where falling drops of water evaporate

stalagmite formed where fallen drops of water evaporate

underground river comes out where limestone rests on clay

## Something to do

There are many crystals you can grow from solution with a little patience and care. An alum is a good material to start with, because it grows easily into well-formed crystals. Directions are given on page 11.

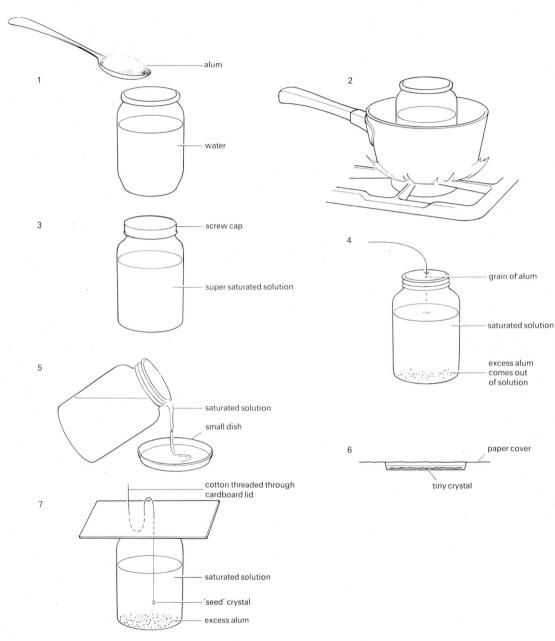

*To grow alum crystals. Don't let the solution warm up while the crystal is in it!*

# TO GROW ALUM CRYSTALS

You can buy potash alum from the chemist. 100 g will be enough to begin with. Put two dessertspoonfuls of alum into a clean half-litre jam jar and fill it nearly full with water. Put the jar into a saucepan of water and heat it to dissolve the alum. (Most substances dissolve more quickly and more completely at higher temperatures.) Stir it till all the solid has dissolved. Pour the hot solution into a perfectly clean dry jar with a screw top or tightly fitting lid, and leave it to cool.

It will become 'super-saturated', which means that it will contain more dissolved alum than is usual at that temperature. Then add a grain or two of alum. Many small crystals will form, leaving a 'saturated' solution, that is one that contains its full amount of dissolved alum at that temperature. Pour off a little of the saturated solution into an egg-cup or small dish and leave it, lightly covered to keep out the dust, for a day or so, depending on the weather. A few small crystals will appear in it. Choose one of them to grow. The smaller it is, the more likely it is to be a single crystal of perfect shape.

Dry your crystal carefully on blotting or clean soft paper or rag. Tie a piece of cotton round it and suspend it in the saturated solution. If you thread the cotton through a circle of cardboard, as shown in the diagram, you will find it easier to get the crystal at the right height. The cotton can be pulled through the lid and will remain in position if the holes are not too big. The cardboard serves to keep dust out of the jar and also to slow down evaporation. If the water evaporates quickly many small crystals will be formed, instead of the one big one for which you are hoping. As the water evaporates, solid alum is deposited on the suspended crystal and it grows larger.

If you are careful and persevering you may grow a fine crystal like the one shown here. You will find more details on crystal growing in the Nuffield chemistry background book *Growing Crystals*.

*Here is a photograph of a single alum crystal, over 3 centimetres across, which is in the Science Museum in London. Choose one of your small crystals which is this shape. You are more likely to find a perfect one among the very small crystals, but it will be difficult to suspend a very tiny one.*

### Some hints

1. Keep your jar in a place which does not change very much in temperature. A basement is the best place; a north room is better than a south-facing one.

11

Better still, stand your jar in a basin of water. The water does not change in temperature as quickly as the air, so the solution round the crystal is protected from sudden temperature changes.

A drop in temperature will only cause more of the solid to be deposited on the crystal, so does not matter. A rise in temperature, on the other hand, will make the solution able to dissolve more of the solid, and your precious crystal will dissolve. If there is any risk of this happening, lift out the crystal, dry it carefully, and only put it back in the solution when the temperature is falling again. It is a good idea to lift it out every morning, if you are out during the day, and replace it in the evening, as the day cools.

2. It is a good idea to keep some extra solid at the bottom of the jar. This reduces the risk of the solution becoming unsaturated and dissolving the crystal if it gets warmer. Alum will be deposited on the solid at the bottom, as well as on the crystal, so its growth will be slower but safer.

3. Tiny crystals will form on the cotton thread as water evaporates from it. Remove these carefully, or they will fall on the crystal and grow there, making it a curious shape.

4. Handle the crystal as little as possible. Even perspiration on the fingers will spoil its faces by dissolving parts of the surface.

*These are alum crystals that have been allowed to grow together. Notice how the angles between the faces remain the same even though the crystals interpenetrate each other, so that none is complete.*

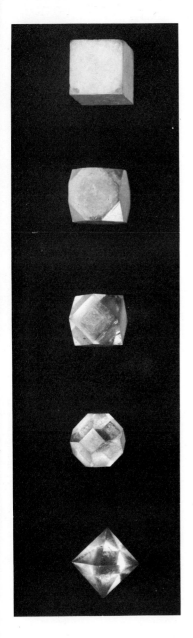

# GEMSTONES & CRYSTAL SYSTEMS

'A diamond is for ever', they say. Certainly, it will last a very long time, for diamond is the hardest material known. Ruby and sapphire are forms of the next hardest substance, aluminium oxide.

Diamond, ruby and sapphire are the most precious of gemstones, and are used in crowns, tiaras, rings and bracelets. Their value depends first of all on their beauty of colour and lustre, but if they were less hard they would soon be damaged, their beauty would not last and they would be less precious.

They often occur in regular, attractive shapes, though their appearance is usually enhanced still more by cutting and polishing.

## CRYSTAL SYSTEMS

Let us look, then, at some of the natural shapes of crystals. You might think each crystal is different from every other, but they can be classified into six main groups or 'systems' as they are called. They are shown in the diagram on the next page. The division of crystals into these systems is based on imaginary lines or axes which can be drawn through their centres. So all crystals through which you can draw three equal axes at right angles are said to belong to the isometric or cubic system. Their measurements are the same in each of these three directions, and so are their properties. The shape, or 'habit' as it is called, of a crystal depends on its system, but several different shapes may belong to the same system.

The alum crystals shown here all belong to the cubic system because all of them have three equal axes at right angles to each other, even though they look so different at first glance. They could all be cut from a cube without cutting off any of the axes. The systems may be divided up again into classes according to their symmetry. If you would like to know more about symmetry, turn to Appendix 1.

*Alum crystals: the top one is a cube, the bottom an octahedron. Those in between show combinations of the two forms.*

13

# THE CRYSTAL SYSTEMS

| name of system | crystal axes | | examples |
|---|---|---|---|

**name of system**

**crystal axes**

**examples**

**cubic or isometric**

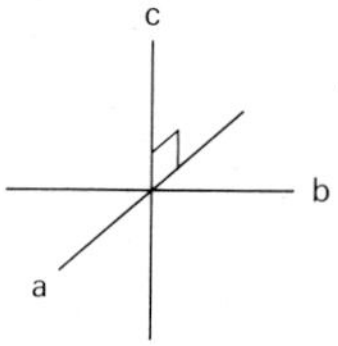

c / b / a

3 equal axes
all at right angles

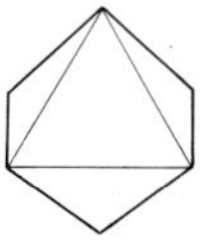

diamond

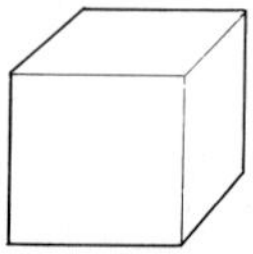

fluorite

**tetragonal**

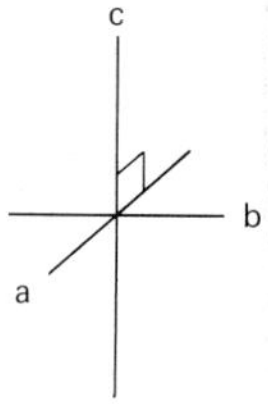

c / b / a

3 axes at right angles,
the vertical one, c, being
longer than a and b

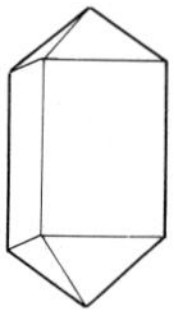

zircon

wulfenite

**hexagonal**

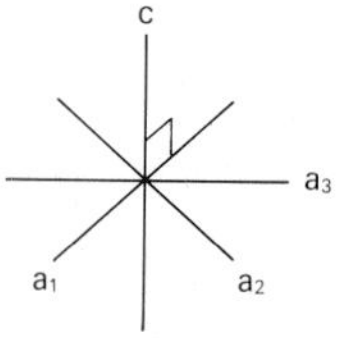

c / $a_3$ / $a_1$ / $a_2$

1 vertical axis
3 equal horizontal axes

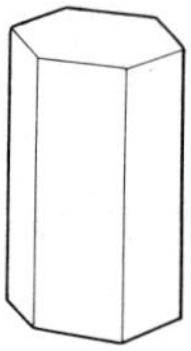

beryl

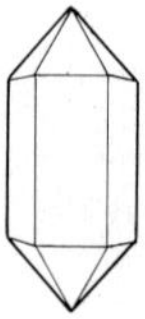

quartz

**orthorhombic**

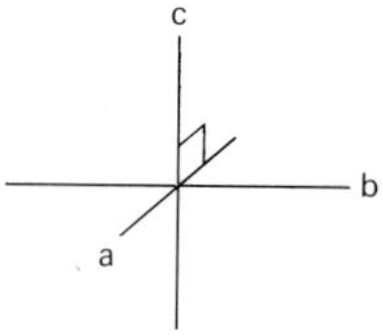

c / b / a

3 axes at right angles
all of unequal lengths

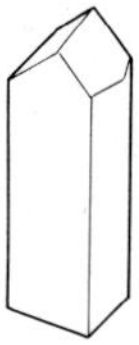

stibnite

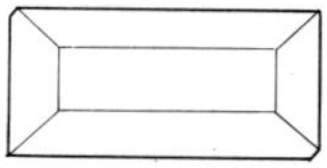

barytes

**monoclinic**

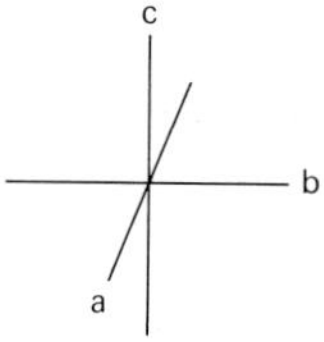

c / b / a

1 vertical axis c
1 horizontal axis b
1 axis a tilted to the others

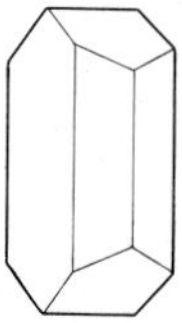

feldspar

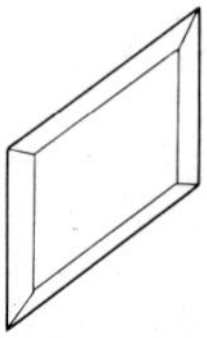

selenite (gypsum)

**triclinic**

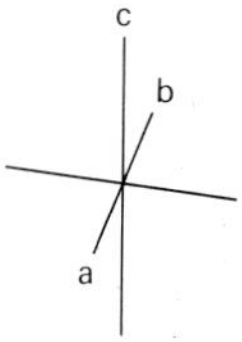

c / b / a

3 unequal axes
tilted at angles which are
not right angles

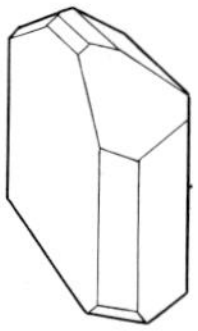

rhodonite

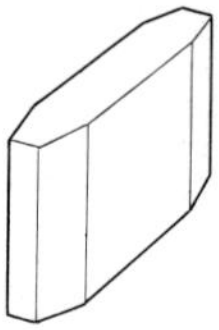

axinite

# 1. Diamond

Diamond belongs to the isometric system since it has three equal axes at right angles to each other. Diamonds were originally formed in volcanic pipes from patches of carbon, under conditions of intense heat and pressure during eruptions billions of years ago. Many diamonds that reached the surface as the rock weathered were carried away by rivers, and they have been found in large numbers as pebbles in river gravels, particularly in India, Brazil and South Africa.

*Right: these diamonds are lying among the grains of rock in which they were found in Siberia. They have been washed down by the river but have not been so badly worn that you cannot see their original habit – an octahedron.*

*Below: the cutting of a diamond.*

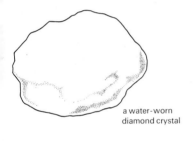

a water-worn
diamond crystal

Although diamond is the world's hardest natural substance (probably ten thousand times more resistant to abrasion than the next hardest substance), a diamond crystal contains planes of weakness along which it can be cleaved, as you have seen calcite cleaved. A skilled craftsman in this way exposes natural faces of the diamond, and afterwards may cut extra surfaces. The reflection and refraction (bending) of light in a properly cut gem diamond is what gives it its characteristic fire.

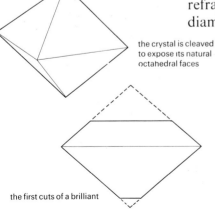

the crystal is cleaved
to expose its natural
octahedral faces

the first cuts of a brilliant

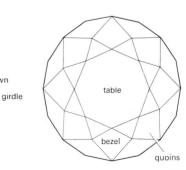

crown

girdle

pavilion

culet

table

bezel

quoins

15

Diamonds have now been made synthetically. For over a hundred years, there have been claims by people who thought (or pretended) they had made them, but it was not until 1953 that diamonds were really synthesised for the first time, by the Swedish ASEA company.

Above: *these diamonds, much enlarged in the picture, were made by De Beers for use on the edge of saws to cut stone or concrete (see the photograph of a saw on page 17). They are made from carbon but cost just as much to make as to mine because very high temperatures and pressures are needed, so there is no hope yet of diamond tiaras appearing in the bargain basement!*

Below: *here a technician is inserting a capsule of graphite into a press. It will be subjected to a pressure of 50 000 kg weight per square centimetre and a temperature of nearly 2000°C. Part of it will be turned into diamond. The capsule containing the graphite is made of aluminium silicate called pyrophylite, and besides graphite also contains a metallic solvent.*

Synthetic diamonds are made by copying nature and applying very high temperatures and pressures to carbon in heavy presses.

To know about diamonds, you can write to the Industrial Diamond Information Bureau, 7 Rolls Buildings, London EC4, who produce free booklets and wall-charts, and you will find some interesting stories in *Crystals, Diamonds and Transistors* by L. W. Marrison (Penguin).

Four-fifths of the world's output of diamond is used in

industry, mainly for cutting, drilling, grinding and polishing. Some diamonds are used whole, set into tips of saws and drills which will then cut and drill the hardest rock.

This saw (*left*) is $2\frac{1}{2}$ metres in diameter, and tipped with diamonds. It is used to cut through layers of quartz in a limestone quarry in Belgium. The edge of the saw has a speed of about 40 metres per second.

These 'dressing tools' are used to produce a fresh surface on a grinding wheel that is clogged or mis-shapen from use. Diamond grinding wheels are often

*Diamond-tipped dressing tools.*

*Diamond-tipped saw.*

used for the last stage of grinding tools whose size must be very accurately controlled. Some surgical instruments, for example, have to be made precisely: they must not measure one five-thousandth of a centimetre too much or too little.

Most industrial diamonds are used as crushed grit bonded into grinding wheels, saws and drills, and are used on plastics, glass, stone, concrete, ceramics, metals and hard metal alloys. The stylus of a gramophone pick-up is often tipped with a diamond.

## 2. Zircon

Zircon, from the tetragonal system, is a semi-precious stone: it is not so beautiful, hard or rare as diamond. The crystals are usually brownish and are formed in some types of igneous rocks. When these are weathered away the hard zircon crystals survive and get carried away in rivers, so that they may be found, like diamonds, in gravel beds.

If they are not too water-worn they may have the habit shown here. They belong to the tetragonal system, which has a longer vertical axis than the cubic system. The natural brownish crystals are heat-treated to make them into more attractive gemstones: red jacinths, blue starlites or pale yellow jargoons. Look at the colour plate.

A*

*Zircon crystal.*

*You can see the hexagonal face of this beryl crystal. Crystals like this are very common in some places.*

## 3. Ruby, sapphire and emerald

These precious stones, which come next in value to diamond, are all examples of *hexagonal* crystals.

Emerald is a variety of a green mineral called beryl.

The finest beryls come from Colombia or from the Ural Mountains. They may be as much as eighteen feet long, but are no use as gems. Real emeralds, which are flawless and deep green, are much more rare. See the colour plates.

Ruby and sapphire are both forms of aluminium oxide, which is colourless. But ruby contains small quantities of chromium oxide and iron oxide, which colours it red; sapphire owes its blue colour to a trace of titanium oxide. Although they belong to the same system their crystal habits are different, so they do not look the same. Corundum, used on emery paper, is an impure form of the same mineral.

Quartz and calcite, which we have already met, also

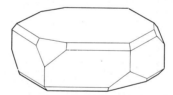

a ruby crystal looks like this if it is perfect . . .

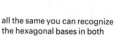

. . . and a sapphire like this . . .

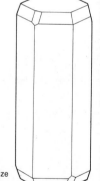

all the same you can recognize the hexagonal bases in both

belong to the hexagonal system. Quartz has many forms, often showing the obvious six-sideness of the hexagonal system. Calcite, too, has many forms, some of them very beautiful, but it is too soft to be used as a gemstone.

## 4. Topaz

Topaz is a gemstone which is an example of the ortho-rhombic system, having three axes, all at right angles but all unequal. It cleaves very easily, so very great care must be taken by stone cutters working it.

*Topaz crystals.*

These crystals are colourless but other colours are found: yellow, pink, bluish and greenish. The wine-yellow stones from Brazil are the most valuable and are called precious topaz. Beautiful sky-blue topazes have been found in the stream beds of the Cairngorms, Scotland.

## 5. Kunzite

The monoclinic system is rather rare in gemstones. Only the fine pink or lilac crystals known as kunzite, and the green hiddenite, are precious stones.

The system has three unequal axes, one of them being tilted, so the crystals of this system have much less symmetry than those of others.

*Orthoclase feldspar.*

Orthoclase feldspar is another monoclinic mineral commonly found in igneous rock.

You might think the pink crystals in the photograph lovely enough to be gemstones, but only an extremely rare variety, a transparent yellow, are precious. The ordinary feldspar is used in the manufacture of porcelain, and is also a source of aluminium.

A bluish opalescent variety is beautiful when polished. It is called moonstone.

## 6. Amazonstone

Amazonstone is a kind of feldspar which has the same composition as the orthoclase feldspar, but crystallises in another system: the triclinic system. Here the crystal axes are all unequal in length, and not at right angles to each other.

*Amazonstone.*

The picture shows some beautiful, well-formed crystals. They are a whitish-green colour. Because the feldspars are very common in granite and because they are not as hard as diamond and ruby, they are not very precious.

### Something to do

Trace the outlines of the figures opposite on to stiff paper or card. Cut them out, fold in the flaps and gum them together to make some models of simple crystal forms. Can you name some crystals they might represent?

# MODELS TO MAKE

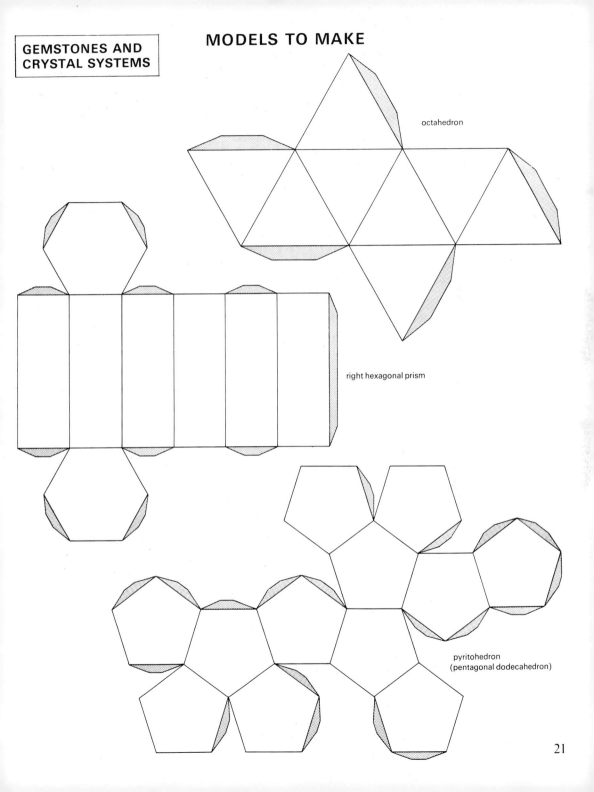

octahedron

right hexagonal prism

pyritohedron
(pentagonal dodecahedron)

# WHICH IS WHICH?

You are not likely to find many gemstones on your holidays, but some common minerals occur as crystals visible to the naked eye and you might well find these if you are observant. How can you recognise them? How do mineralogists and crystallographers tell which is which?

The last chapter told you something about crystal form, which is a help if you want to identify a crystal. But although each substance has a particular habit (that is, a shape when fully developed), some substances, like calcite, crystallise in several different forms. Also there are more substances than habits, so a particular shape of crystal may be found in several minerals. This means that we must look at other properties.

## COLOUR

Colour is probably the first thing you notice about a crystal, but it is not always a reliable guide. You would not be likely to mistake the bright orange of potassium dichromate (colour plate), or the blue fluorite in its very obvious cubes (colour plate). On the other hand quartz may be found not only clear, but also rose, blue, yellow, amethyst, smoky or cloudy, and colour is then a risky guide to its identity.

The streak made by a substance is a safer guide. This is really the colour of its powder. A tiny scratch can be made on a surface of the mineral, or the mineral itself can be rubbed on a piece of unglazed china or ground glass. See Appendix 3.

### Flame test

The colour that a substance gives in a flame is a quite certain indication of certain chemicals in it. For example, any compound of sodium will colour a flame bright yellow; strontium (from, say, the mineral celestine) gives the splendid crimson red that is responsible for the colour of some fireworks; copper produces a bright green, and potassium colours the flame lilac.

**Something to do**

Kitchen salt is sodium chloride. Drop a little salt into the flame of a gas ring to see the bright-yellow sodium light. Washing soda is sodium carbonate. Try a little of that. Do you get the same colour? Try a few tiny crystals of your alum. What do you think is in the alum, judging by this test? Hold the end of a piece of thin copper wire in the flame. You will notice the characteristic green flame that identifies copper, but you will also notice that you have to put the wire down suddenly. What does this tell you about copper?

# LUSTRE

Some crystals have a shine or lustre like a metal. Some of these are indeed metallic ores, like these crystals of pyrite.

*Pyrite.*

Pyrite is iron sulphide, but it is such a bright shiny yellow that it is often called 'fool's gold'. It is quite a common mineral. You will almost certainly see it if you look carefully at lumps of coal. Have you been caught already, and thought you had found gold in your coal-scuttle? Can you see some cubic crystals in the photo? Can you see some faces of crystals which are the five-sided figure called a pentagon?

A regular solid which has twelve faces, each being a regular pentagon, is called a *pyritohedron*. Can you guess why? Notice also the fine lines called striations on many of the faces. They are the result of what might be called some indecision on the part of the crystal as to whether it will become a cube or a pyritohedron or a combination of the two.

Some crystals which are not metallic show this kind of lustre. Graphite, for example, which you know as pencil 'lead', has a metallic shine, although it is a form of carbon which is not a metal. Quartz and calcite are said to have a *vitreous lustre*: they look *glassy*. Other crystals may be *pearly*; dolomitic marble is called pearl spar for this reason. Satin spar is a variety of gypsum. The long thin monoclinic crystals are all aligned the same way and shine like satin. You can see a layer of pale-pink satin spar, like cream in a layer cake, in the rocks close to the Severn Bridge at Aust Cliff.

## TRANSPARENT CRYSTALS

Some forms of quartz and calcite allow light to pass and you can see through them. They are *transparent*. Other crystals transmit no light and are *opaque* like amazonstone; many come in between. Still others allow light to pass but you cannot see through them. They are *translucent*. (Frosted glass is translucent but it is not crystalline.)

Rays of light are bent (i.e. refracted) by all transparent crystals as they enter or leave the surfaces. All crystals except those of the isometric system bend them differently in different directions. This results in one ray of light being split into two as it passes through the crystal.

*Calcite.*

This effect is called double refraction and is very obvious in calcite. If you look through the crystal you can see two images of a letter beneath, as is clear in this photograph of a calcite crystal.

### Something to do

1. Stand a pencil in a glass of water so that it is leaning sideways. Look at it from the top. Do you think the pencil has got bent? Run your finger along it to make sure. Well, what is bent then? This phenomenon is caused by refraction. Fill a medicine bottle with water and look at the pencil through it. Does the part of the pencil you can see above the bottle seem to be in line with the part you can see through the water? What is happening?

look at the pencil through the side of the bottle, putting your eye first nearer the bottom of the bottle, then at the middle, then nearer the neck end

pencil

bottle full of water

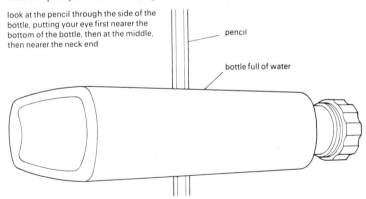

2. Polaroid sunglasses are often sold with a small circle of polaroid in a cardboard mount. An optician will save a few of these, if you ask, or perhaps you could borrow somebody's polaroid sunglasses. Find a patch of well-polished surface near a window or an electric light, and stand so that the light reflected from it makes a glare in your eyes. Look at it through the piece of polaroid. If you rotate the polaroid, you can find a position in which there is hardly any glare.

## POLARISED LIGHT

Polaroid is a thin transparent plastic in which many tiny crystals are lined up all in the same direction. They split the light into two rays, just as calcite does, but then they absorb one of the rays. The ray that is not absorbed and gets through is *polarised* light. It is called polarised because the crystals have forced the waves of light to vibrate in, say, an up-and-down direction only. Ordinary light waves vibrate not only up and down but left to right, skew-left and skew-right and every way. The waves of polarised light vibrate in only one plane.

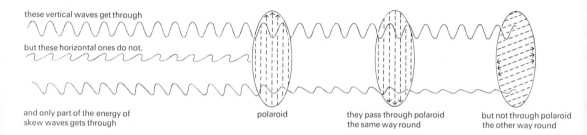

these vertical waves get through

but these horizontal ones do not,

and only part of the energy of skew waves gets through

polaroid

they pass through polaroid the same way round

but not through polaroid the other way round

The waves of 'glare' light are partly polarised, so a polarised filter in a particular position will stop most of them. You will understand more of this in later physics lessons when you have found out how light behaves. It is mentioned here because this ability to polarise light is a property of most crystals. It can be used to identify minerals, to save the eyes from glare by polaroid sunglasses and in other ways. Look at the colour plate, which shows a spanner being tested between crossed-polaroids.

### Something to do

Make a concentrated sugar solution by stirring up sugar in water till no more will dissolve. (Yes, sugar is crystalline; look at the tiny crystals under a magnifier.)

Pour the solution into a clear glass tumbler or, better, a medicine bottle. Now arrange two circles of polaroid so that the first one polarises the light, and the second one does not let it through (see the diagram on the next page).

(You won't see any dotted lines on the polaroid of course. You will have to turn the second one round till, in fact, no light comes through and they look black, or nearly black.)

When you have fixed them in this position, put the bottle of sugar solution between the 'crossed' polaroids, with one of the narrow sides facing you so that the light has the longest path through it.

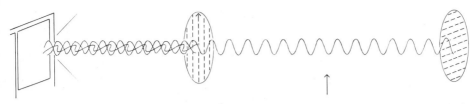

| window | first polaroid | bottle of solution in here | second polaroid | your eye |

What do you notice?

Can you alter the position of the second polaroid so that the light is again cut off?

You will find this easier to manage if you fix your second piece of polaroid across the end of a short length of cardboard tube, so that you can rotate it easily. (Don't try to hold everything in your hands!) You have really made a simple *polarimeter*. Try some pieces of cellophane and transparent plastic or sellotape instead of the sugar solution. Fold the cellophane so that light passes through different thicknesses. Try twisting the plastic while you are looking at it. You may find it easier to invent a polarimeter in which the polaroids are horizontal, one above and one below a tube or tumbler full of the sugar solution.

## HARDNESS

It is easy to test the hardness of crystals by seeing how easily they can be scratched. (Choose a tiny spot that will show least.) The ten substances below have been chosen to make a scale of hardness, working up from the mineral talc, the softest, to diamond:

| HARDNESS | MINERAL | ROUGH TEST |
|---|---|---|
| 1 | talc | soft, slippery feel (talc is the chief ingredient of talcum powder) |
| 2 | gypsum | can be scratched by a fingernail |
| 3 | calcite | just scratched by a copper coin |
| 4 | fluorite | scratched by a knife without difficulty |
| 5 | apatite | scratched by a knife with difficulty |
| 6 | orthoclase | scratched by a file and will scratch glass |
| 7 | quartz | scratches glass easily |
| 8 | topaz | scratched by emery paper, will scratch glass very easily |
| 9 | corundum | scratched only by diamond |
| 10 | diamond | the hardest known substance |

Each mineral will scratch those having a lower number in the scale. The scale is called after the Austrian scientist Mohs, who put it forward in 1822.

27

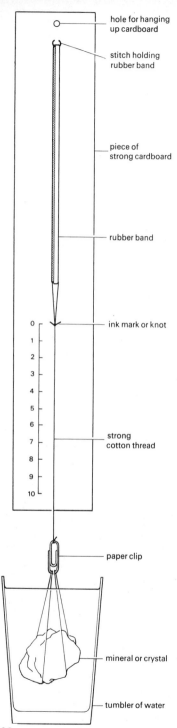

hole for hanging up cardboard

stitch holding rubber band

piece of strong cardboard

rubber band

ink mark or knot

strong cotton thread

paper clip

mineral or crystal

tumbler of water

## Something to do

Collect as many different small stones from a path, or pebbles from a beach, as you can and make your own hardness scale.

Your fingernail will scratch number 2 in Mohs' scale, but not number 3, so a fingernail must have a hardness of about $2\frac{1}{2}$, and it will scratch anything less hard. Similarly a steel knife must have a hardness about $5\frac{1}{2}$, and anything it scratches must have a hardness less than $5\frac{1}{2}$. Emery paper is made of corundum powder stuck on paper, so you can easily get number 9 in your scale. Very few minerals are as hard as, or harder than, quartz, and these are mostly precious stones which you will not be needing to test! When you have found a set of pebbles that are approximately of hardness 1–8, with emery as 9, you will be able to use them to test other materials such as building stone.

## A problem to think about

You have two colourless crystals. They are both hexagonal and you are told that they are calcite and quartz. How could you tell which is which? If you are beginning to study chemistry, you should be able to think of another way, as well.

# SPECIFIC GRAVITY

A lump of barytes, say, which is an ore of barium, feels much heavier than a lump of calcite of the same size. We can express this quality most easily by comparing all substances with water. If we say, for example, that the *specific gravity* of ruby is 4, we mean that a ruby is four times as heavy as a 'lump' of water the same size; pyrite, having a specific gravity of 5, is five times as heavy as an equal volume of water.

## Something to do

Make yourself a simple specific gravity balance and use it to find the specific gravity of minerals, pebbles, coal, etc.

Choose a rubber band which stretches a few inches when you hang your specimens from it. Fix it to a piece of strong cardboard (a stitch through both would do) and hang a *long* loop of strong cotton from the bottom of it. A knot or an ink mark at the top of the cotton will serve as a pointer. A paper clip on the end makes it easier to suspend your specimen. Mark a scale starting from zero at the knot. It does not matter what the divisions are as long as they are all equal.

To find the specific gravity of a crystal tie thin cotton round the crystal and hang it from the clip.

Read the pointer – suppose it says 10 divisions. Now lower the crystal gently and completely into water. See that it is hanging freely and not touching the sides of the glass.

Read the pointer again. Suppose it says 6 divisions. Now you can work out the specific gravity. It is given by the weight of the specimen divided by its loss of weight in water.

In this case, $\dfrac{10 \text{ divisions}}{(10-6) \text{ divisions}} = \dfrac{10}{4} = 2\cdot5$

So this crystal is two and a half times as heavy as an equal volume of water.

If you want to know why this method gives you the relative density of the crystal, look up the chapter on Archimedes' principle in a physics textbook.

*Two warnings*

1. Be careful not to use too thin a rubber band, or it may not stretch equally as you load it equally. See problem 1 below.
2. Do not risk the precious alum crystal that you have grown. Alum is soluble in water, so this is not a good method for finding its relative density!

*And the problems*

1. How could you test your rubber band to make sure that equal divisions on the scale really do stand for equal increases of weight on it? In other words, is the stretch proportional to the load you hang on the band?

*Hint:* Your mother would probably lend you curtain rings or hooks, or press fasteners; or you could perhaps borrow screws or sixpences from your father.

2. (*Difficult*) Can you think how to find the relative density of your alum? It does not dissolve in oil, but the statement you used about water is not true for oil. You could borrow a measuring cylinder if you liked.

*Note:* You will find a useful table of hardness and relative density in Appendix 3.

# CLEAVAGE

Cleavage is the tendency of some crystals to split along certain planes. If a crystal is broken it may split much more easily along certain surfaces which are called cleavage planes. You have seen calcite cleaved in your lessons, and have noticed the smooth surface that is exposed.

*Mica crystal.*

Mica cleaves very easily indeed in one plane in the same direction as the base of the crystal, so easily in fact that it seems to be made up of a pile of very thin plates, as you can see in the photograph of a muscovite mica crystal on the previous page. It gets its name from Muscovy, the old name for Russia, where it was an early form of window glass. It is a very good electric insulator and you may be able to get a piece from a broken electric iron or an old radio valve. You can split it, almost indefinitely, into thinner sheets. You might guess by looking at this six-sided crystal that it belongs to the hexagonal system. Actually it is a monoclinic crystal; one axis is inclined at an angle which is not a right angle.

*Galena, a common form of lead ore, splits easily in three directions at right angles to each other. If it is struck it breaks into little rectangular bricks, as you can see here. It is said to show cubic cleavage.*

## OTHER CLUES

Some crystals have other special properties which help in their identification. Magnetite (an ore of iron) is, as you might guess from its name, magnetic. There are old sailors' yarns about magnetic rock called lodestone, which pulled all the iron nails out of wooden ships, bringing them to their doom! The tales are exaggerated but the natural magnetic rock does exist. A few other metallic crystals show a weaker magnetism, for example those of nickel and chromium.

Uranium and thorium compounds are radioactive, so they can be identified by the 'clicks' they produce when a Geiger counter is brought near them. But it is only occasionally that uranium ore occurs in a form that is visibly crystalline. When it does, the crystals are isometric and show faces of the octahedron.

Some crystals fluoresce when ultra-violet light is shone on them and show colours quite unlike their usual 'day-light' colours. Look at the colour plate of franklinite ore.

These and other properties, taken together, make it possible to know one kind of crystal from another.

# THE INSIDE STORY

You will realise by now that there is a certain order and regularity about crystals. Let us have a look at this regularity and see if we can guess what might be behind it.

## ANGLES BETWEEN CRYSTAL FACES

First, every crystal has a definite shape. When a substance crystallises all the crystals are of one kind, or sometimes of just a few related kinds. Often they have no room to grow into a definite shape, but, even so, it is always true that the angle between next-door faces is the same size every time.

*Look at these copper sulphate crystals which have grown from a thin layer of solution on a glass slide. Notice that the angles between the faces are always the same, even when the crystal has not had 'elbow room' to grow.*

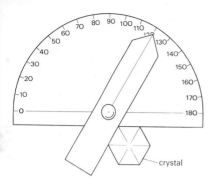

crystal

### Something to do

Make yourself a simple goniometer so that you can measure the angles between the faces of crystals.

You will need two pieces of cardboard or, better still, transparent plastic. Trace the markings of a protractor on to the card. Cut a straight piece to a point at one end. Make a hole through the centre of this piece and through the centre of your circular scale. Use a paper fastener to hold them together, but not so tightly that you cannot rotate the arm.

Put the crystal between the pointer and the edge of the scale as it is shown above and you can read off the angle between the two faces from the pointer.

## CLEAVAGE

We mentioned on page 29 the tendency some crystals have to split along certain planes, i.e. to show cleavage. When crystals break in this way, the small pieces which are formed are related in shape to the original crystal.

Here, for example, is a photo of a cleaved calcite crystal.

PLATE 1
*Potassium dichromate crystals*

PLATE 2
*Fluorite crystals*

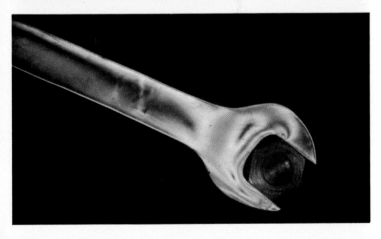

PLATE 3
*A model spanner made in clear plastic and photographed between two sheets of polaroid. The coloured bands are called photoelastic stress patterns, and show up weaknesses of design*

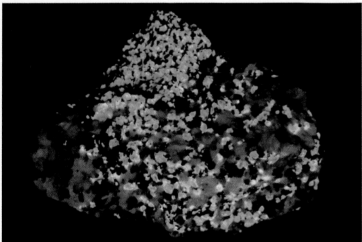

PLATE 6

*A piece of franklinite seen (a) in ordinary light and (b) in ultraviolet light*

*The black crystals are franklinite, an oxide of iron, zinc and manganese. The white crystals are calcite, which becomes bright red in ultraviolet light. The brownish crystals are willemite, zinc silicate, which fluoresces green*

PLATE 7

*A tourmaline crystal embedded in quartz. Tourmaline has been valued as a gemstone for many centuries, but also became valuable during World War II for use in pressure gauges (see page 48)*

A whole crystal of calcite is a shape called rhombo-hedral. The angles at the corners are all 78° or 102°. When it is broken the pieces also have their angles these same sizes, and are also rhombohedrons. The mica on page 29 cleaves into thin plates with surfaces like those of the large crystal; galena, a cubic crystal, into smaller cuboids, as shown on page 30. Now we begin to wonder what could lie behind this behaviour.

## ETCHING

Again, if we etch a crystal we can see something more of this small-scale regularity. To etch a surface we let some liquid (or gas) that will 'eat away' the substance chemic-ally, act on the surface for a short time, then wash it off and look at the surface under a microscope.

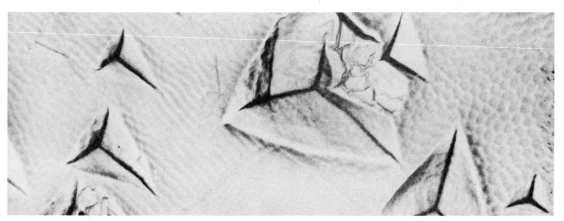

The photograph shows aluminium which has been etched by acid. Many small crystals can be seen, all with tetrahedral tops (like three-cornered hats) and their ridges all lined up in the same direction. This photograph has been taken with an extremely powerful kind of micro-scope. The real crystals are fifteen thousand times smaller than they appear here. The process of etching re-moves the softest or weakest parts of a surface, leaving little ditches or pits which trace out the structure underneath.

Fluids used for etching are very corrosive: they will eat away many substances besides the one you are etching.

So this is not something you can do yourself; you will have to be content with seeing etched tin crystals in the laboratory.

## WHAT IS UNDERNEATH?

Constant angles, cleavage, etching patterns – we can make sense of all of these if we imagine that crystals are made of very tiny 'building bricks' all of the same size and shape, arranged in a regular pattern.

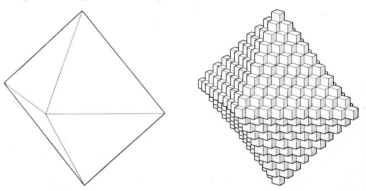

The shape of this alum crystal might well be accounted for by assuming that it is composed of tiny cubes or spheres built up like a pile of oranges (above). The smaller the oranges, of course, compared with the size of the pile, the less one would notice the bumps: the surfaces would appear smooth.

Some alum crystals are cubic and they could be made of a pile of bricks like this:

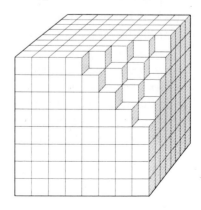

One corner is missing. So the cube has one face of the octahedron shown in the diagram above it.

Nowadays, after long years of careful observation, with brilliant, inspired guessing and repeated painstaking experiments which you will hear about later, we think of the building blocks of crystals as atoms; or electrically charged atoms (we call these 'ions') that push and pull each other; or groups of atoms called molecules.

## CRYSTAL LATTICES

Common salt is a compound of sodium and chlorine, so its chemical name is sodium chloride. The tiny cubes which are the crystals of common salt can be thought of as built up of equal numbers of sodium and chloride ions arranged in a regular pattern like this.

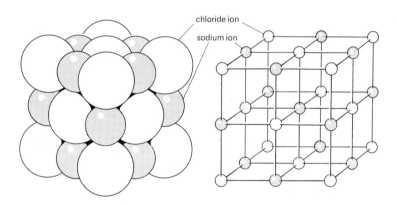

chloride ion

sodium ion

It is clearer if we show the positions of the ions by smaller balls as shown in the second drawing.

The black lines joining the balls represent the forces with which the ions attract each other. You must not think of them as sticks or rods, any more than atoms are really like beads or balls. The pictures just show simple models that we can have in our heads to help us think about the behaviour of crystals. The regular repeating pattern we call a crystal lattice. So we can explain the cubic shape of crystals of common salt by supposing that there is a simple cubic lattice of ions inside it.

In a similar way, we suppose that the crystal lattice of a calcite crystal is a rhombohedral arrangement of ions. Compare the shape of the rhombohedral calcite crystals shown on pages 25 and 32 with that of the crystal lattice shown below.

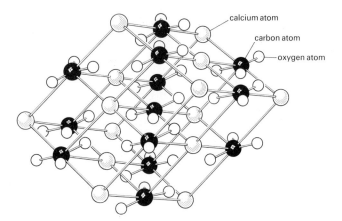

calcium atom

carbon atom

oxygen atom

*This diagram shows the crystal lattice of calcite which is calcium carbonate. Ions of calcium, carbon and oxygen form a lattice which is rhombohedral, that is, shaped like a cube squashed sideways.*

If you would like to know more about this, turn to Appendix 2 where you will find diagrams of all fourteen possible lattices. They are enough, between them, to explain all the crystal shapes we find in the six crystal systems.

## ARRANGEMENT AND BEHAVIOUR

The arrangement of atoms makes a very big difference to the properties of the substance. The precious hard diamonds and soft black graphite of a pencil 'lead' are both made of pure carbon, and it is the different lattices that make their properties so different.

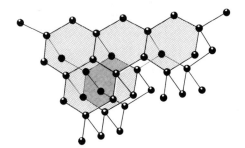

*In diamond, each carbon atom is joined by four equally spaced 'bonds' (forces of attraction) to other atoms, so the whole structure is very closely knit together, and the substance is very hard.*

In graphite, the carbon atoms are arranged in flat sheets, which are too far apart to have much effect on each other. Each carbon atom is only attached to three others, in a hexagonal lattice, but these bonds are strong; some of them are double bonds.

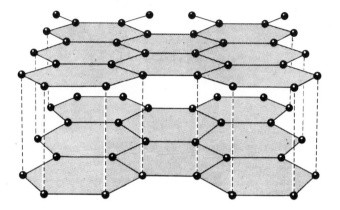

It is easy to see why graphite cleaves easily into very thin plates or leaves, and why it is so different from diamond, although it is made of the same material.

### Something to do

Have you a drawer that does not slide in and out easily? Try rubbing your pencil 'lead' on the edges of the wood. Does this improve its movement? Graphite can be used as a lubricant because the sheets of atoms slide easily over one another, making movement between surfaces less difficult. If you have no jamming drawers, rub your pencil on a piece of wood, and feel the difference with your finger.

Perhaps you have seen someone ease the stem of his pipe by lubricating it with graphite from his pencil? Watch out for this little habit.

# CRYSTALS FOR TODAY AND TOMORROW

Crystals are far more common than you might suppose. Very many of the everyday materials around you are crystals. In fact, all true solids are crystalline, that is, they have their atoms arranged in a regular pattern.

Later on in your physics course you will learn how physicists, by shining beams of X-rays on crystals, are able to produce patterns on the film of a camera. From these they can work out the spacing and arrangement of atoms in a crystal. Sir Lawrence Bragg in Cambridge and Dame Kathleen Lonsdale, an eminent woman scientist, have done brilliant work in this field.

Some hard, rigid materials like glass and plastics you might think of as solids, but they are not. Their atoms are irregularly arranged, and in some ways their behaviour is more like that of a liquid. They have, for example, no particular melting temperature. Ice, which is crystalline, and so is a true solid, melts at 0°C (unless it is under pressure), whereas glass just gets softer and more liquid as you make it hotter. It is not a true solid.

## NATURAL CRYSTALS

Some substances occur naturally as recognisable crystals in the rocks of the earth's crust. These include precious stones, common salt, Chile saltpetre for fertiliser and many other useful materials. But among the most beautiful and perfectly formed natural crystals are some you can easily see for yourself most winters. They are the crystals that fall as snowflakes.

### Something to do

Next time it snows, catch some snowflakes on your coat sleeve, or on a piece of rough cloth; black velvet is ideal. At first the material will be warm and the flakes will melt before you have time to look at them. But be patient for a little time, and when dry snow is falling on a cold surface you will be able to see these lovely shapes for yourself through a magnifier. What do you notice about the shape of snowflakes? You will find a wide variety of different patterns, but all of them are six-sided. What kind of pattern of 'building bricks' would you invent to explain their appearance?

*Ice crystals.*

*More ice crystals.*

## INSIDE A SNOWFLAKE

Below is a picture of a model showing how the molecules in ice might be arranged. The rods connecting the balls represent bonds (forces of attraction, you will remember) between hydrogen atoms.

When ice turns to water this well-arranged lattice breaks down, the bonds are continually 'changing partners' and one might see, at any instant, numbers of free molecules moving about inside a lattice of bonded molecules and rapidly changing places with them. So, thinking of this, we can begin to see why ice and water are so different in their behaviour, although they are made of exactly the same 'stuff'.

# CRYSTALS ROUND THE HOUSE

### Something to do

How many examples of crystals can you find in your own home? Take your magnifier and a notebook. You will find more than you expect. The kitchen is a good place to start.

## COMMON SALT

You will have looked at common salt crystals under a microscope.

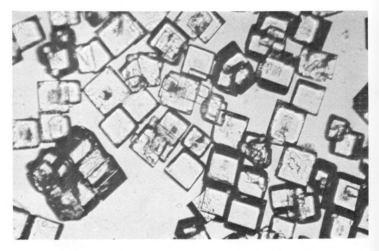

*Here is a picture to remind you. These cubic crystals have been made by evaporating pure brine.*

You know the use of salt in food, and the body's need of it. Many tons are produced annually for this purpose, but many more are needed each year in chemical industries.

If salt is stored for a time it tends to 'cake', and then much time and labour have to be wasted in making it manageable again. You can imagine that it would be dangerous to attack piles of caked explosives with picks or drills to break it up. Caked fertilisers are not usually dangerous, but they can be handled more easily and economically if they consist of small, non-sticking crystals that can be poured, like sand in an egg timer. Nowadays, the chemical industries have found a way of doing this, not only for salt, but for other chemicals also. The trick is to add a little of another chemical to the salt solution before it is allowed to crystallise. The right chemical has to be found to alter the habit of the crystal, that is, to make it grow into a shape which is not its usual one. So salt, which normally has a cubic habit, can be persuaded to crystallise like little trees: i.e. with a dendritic habit. These dendritic crystals are much less likely to cake together than are the cubic ones, so piles of salt can be stored without trouble.

*Dendritic crystals of common salt (sodium chloride). A trace of potassium ferrocyanide was added to make the salt crystallise in this form.*

### Something to find out

You may have seen big piles of salt at the service areas of motorways. It is used in winter to de-ice roads. What does salt do to de-ice roads?

### Something to do

Get some ice from the refrigerator or from a fishmonger, crush it into small pieces and put it in a small basin. Add a handful of salt and stir it. You have made a *freezing mixture*. It is colder than melting ice. If you can borrow a thermometer you can tell how many degrees below freezing point it is. You can try to freeze some fruit juice or custard in a small container by putting it into your ice–salt mixture.

Since a mixture of salt and water has a lower freezing point than pure water, the air temperature will have to be very much colder than 0°C before salty water on the road will freeze.

41

# SUGAR

Sugar is sure to be included in your lists of crystals. But there is more than one kind of sugar. If you looked at icing sugar using a magnifier, you would hardly be able to tell it is crystalline, because the grains are so small. It sets very readily when moistened because such tiny crystals both dissolve and then recrystallise rapidly. The sugar in chocolate is ground so fine that it tastes perfectly smooth on the tongue, yet a microscope will show it as tiny broken crystals.

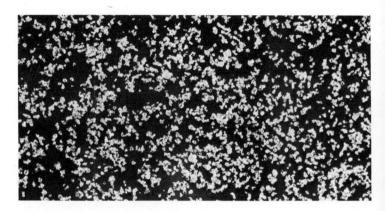

In caster sugar (*above*) you can see the crystals. It is not possible from the photograph to tell which is their crystal system. Actually they are monoclinic.

Granulated sugar (*above*) has crystals larger than those of caster sugar. It dissolves more slowly.

Coffee sugar crystals' (*above*) are larger than those of granulated sugar and are preferred for coffee because they dissolve slowly. The coffee connoisseur prefers to begin with the bitter taste of unsweetened coffee, and then enjoys its gradual sweetening as the crystals slowly dissolve.

### Something to think about

Why do larger crystals dissolve more slowly than small ones? *Hint:* Why does a litre of boiling water cool more quickly in two half-litre jugs than in a litre jug of similar shape?

Did you know that sugar crystals act as tiny knives cutting up the white of an egg when it is beaten?

## OTHER CRYSTALS IN THE HOUSE

Do not forget to keep your eyes open for metal crystals. Brass door knobs that have been etched by long and frequent contact with the sweat on human hands reveal their crystalline structure: you can see the boundaries of the crystals on the surface. If you find broken metal, examine the broken surfaces for crystals. They are also visible on a galvanised tank, bucket or dustbin. These are crystals of the zinc with which the iron has been coated to protect it from rust. The insides of cans that have held tinned fruit or vegetables also show crystals. You may be lucky in the bathroom too: bath salts, Epsom-salt, or Glauber's salt may be there.

You could try growing these, as you grew an alum crystal.

## MAN-MADE CRYSTALS

Attempts to make artificial materials to replace natural ones have been going on for a very long time. Sometimes the natural material is scarce and valuable and a cheaper substitute is looked for. This is the case with diamonds and other stones used for decoration. Or sometimes the natural material is imperfect in some way, and it is hoped that a man-made material will be an improvement. This was the case with quartz crystals, which were synthesised more than a hundred years ago. The crystalline quartz will not dissolve in water, but silicon dioxide (the same material) ground up finely is very slightly soluble in hot water. If a seed crystal is put in this solution, a perfect quartz crystal can be grown.

But why is someone prepared to take such trouble to grow a perfect quartz crystal? Because quartz crystals can be used as time-keepers.

## CRYSTAL CLOCKS

That sounds very odd. How could a crystal possibly keep time? It beats time, if you like, by vibrating to and fro at a very steady rate. The number of to–fro motions each second is called the frequency, and the crystal has its own natural frequency.

You know that a child's swing has its own particular rate of swing which you can alter only by holding on to the seat and forcing it to go at your own chosen frequency. But this is an exhausting and pointless occupation. (Try, if you don't believe it!) In the ordinary way you feed in the energy but let the swing itself decide on the frequency. In other words, you adjust the time of your pushes to match the frequency of the swing if you want it to swing high.

In the same way a quartz crystal has a natural frequency which depends on its size and shape and how 'springy' it is. Electrical energy is fed into it from a circuit which is arranged, 'tuned', we say, to vibrate at the same rate as the crystal, and so the vibrations* of the crystal are kept going.

*Vibrate. *We more often use the word 'oscillate' for vibrations that we cannot see.*

*A quartz-crystal clock at the National Physical Laboratory.*

The frequency may be very high indeed. Quartz crystals can vibrate a hundred million times a second. This is much too fast to see or hear. An electronic circuit can, however, count these oscillations, and move some mechanical device to indicate fractions of a second. And there is your clock.

Quartz-crystal clocks are compared with the motion of the stars which is our long-term time-keeper and are used to control time-signals which are transmitted by radio to give the correct time all over the world.

### Something to think about

How fast do radio waves travel? Is a time-signal from London late by the time it arrives in New York?

45

It is sometimes not necessary to count the vibrations: quartz-crystal oscillators are used by radio transmission stations to 'beat time' so that the wave that carries the B.B.C. Radio 2 programme, for example, continues to oscillate at two hundred thousand cycles a second. If it wandered off that rate we should have to be constantly re-tuning our radio sets to pick it up.

## SYNTHETIC CRYSTALS FOR LENSES AND PRISMS

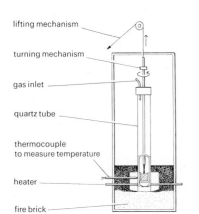

lifting mechanism

turning mechanism

gas inlet

quartz tube

thermocouple
to measure temperature

heater

fire brick

In 1902 Verneuil, a Frenchman, invented a process for making synthetic rubies and sapphires. They were wanted not as jewellery, but to replace glass in certain scientific instruments. The natural gems have too many flaws; the man-made ones are better.

Natural rock salt was used in instruments for studying infra-red radiation – the kind of rays which you get from a heat-treatment lamp. These rays are not transmitted through glass, and so lenses, for example in infra-red cameras, have to be made of clear rock-salt. Large, perfect crystals for this purpose can now be made synthetically.

### Something to think about
Why do glass fire-screens protect your legs from scorching but do not prevent you from seeing the fire?

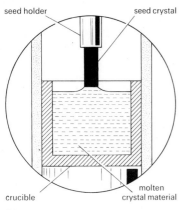

seed holder

seed crystal

crucible

molten
crystal material

Verneuil's process for growing crystals is rather like the way you obtained salol crystals. The substance is melted in a furnace and a tiny seed at the end of a metal holder is dipped into it. As the molten material solidifies on the top of the seed crystal, it is slowly drawn out of the furnace.

The crystal continues to grow on the seed, and finally a large single crystal, perhaps more than 10 centimetres long, is produced.

# ENERGY CHANGES

Energy is something you will go on learning about all through your physics course. But you probably realise already that you need 'breakfast energy' to do work. Electrical energy will also do jobs: carpet cleaning, lifting customers to the fourth floor, spin drying and much else. Chemical energy from fuel, energy of a moving object, energy of heat and light and nuclear energy – all of them are available to do work for us. And each time we see a job being done one kind of energy turns into another kind. The chemical energy of petrol in a car, for example, turns into energy of motion, as the car accelerates.

Devices which are deliberately constructed to exchange one kind of energy for another are called *transducers*. Some crystals can be very useful in this way.

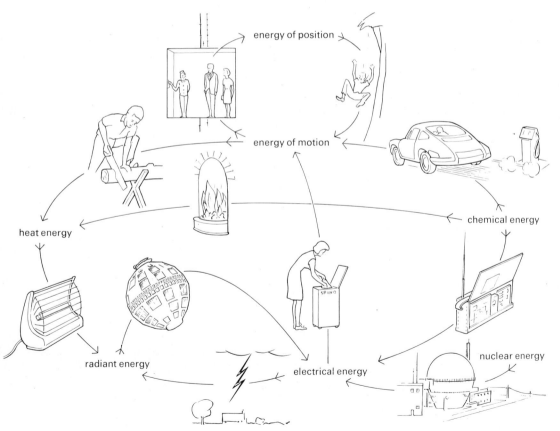

energy of position

energy of motion

chemical energy

heat energy

nuclear energy

radiant energy

electrical energy

## SQUEEZING CRYSTALS

In the 1940s a substance called ammonium di-hydrogen phosphate was grown as single crystals from solution, in much the same way as you grew alum crystals. It has, in common with some other crystals, the property of being able to produce electricity under a changing pressure. This is called *piezo-electricity* – from the Greek word for pressure. If you give a crystal of this substance a smart

*This tourmaline crystal (which is embedded in quartz) also shows the piezo-electric effect. If it is compressed from top to bottom a measurable voltage is set up across it.*

blow with a hammer it produces an electric voltage, that is, a difference of electrical pressure, across its two ends. This voltage can be used to move an indicator of some kind to register the changing pressure. Some pressure gauges work on this principle. In other words, the crystal will change the energy of motion that you supply through the hammer into electrical energy. You might guess that the reverse change could also take place and it does: if you put a voltage across the crystal it becomes distorted. The effect is very small but it can be made very useful. When you have learnt more about electric charges you will be able to think out why these exchanges occur.

# CRYSTAL WIRELESS SETS

Early wireless receivers were often called 'cat's whisker sets. The cat's whisker was a fine wire which just touched a crystal of galena, which is lead sulphide. If this contact were carefully adjusted, it could just manage to pick up a weak signal from an aerial and convert it into a current in an earphone. It was a very doubtful mechanism, and needed much twiddling and patience. So people were very glad when radio valves were invented to do the job more efficiently and reliably. But that was not the end of the story. Crystals were to make a great come-back.

# POCKET RADIOS AND TRANSISTORS

You know that plastic-covered flex carries electricity to your table-lamp or electric iron. The copper wire inside the plastic conducts the electricity and plastic does not, so the electric current flows in the copper and does not make a getaway through your hand when you touch the plastic-covered flex. Copper and other substances like it are called *conductors*; plastics, rubber, porcelain and materials which behave similarly are *non-conductors* or *insulators*. There is, however, an in-between class of sub-stance called *semi-conductors*. These had been known and used since the 1920s, but it was not until 1946 that some serious research work was done on them. Three men at the Bell Telephone Laboratories in New Jersey, U.S.A., set about this job. Brattain, Shockly and Bardeen were their names. They worked out how a single crystal of a highly pure semi-conductor might be made to control an electric current, figuring it out on paper and trying it out in the laboratory. Out of a lot of hard work and failure, inspired guessing, careful thinking and painstaking experiment, was born the *transistor* in June 1948.

Semi-conductor materials of which transistors are made must contain less than one part in a thousand million of impurities, and very small, precise quantities of other materials are added. A transistor can be thought of as a three-layer crystal sandwich.

49

A typical transistor is very small, not much bigger than a full stop, and even in its outside case not as big as a centimetre length of pencil. Yet it can control currents much bigger than the one it carries, just as the small force of a man's foot can control the large force that accelerates a motor car.

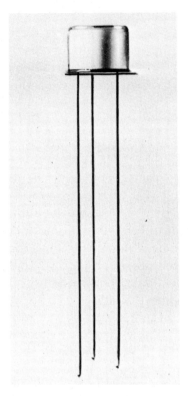

Above: *a transistor inside its capsule. It is connected in an electrical circuit by the three long wires.*

The development of transistors has made it possible to produce pocket radio receivers like the little one in the photograph, and they can be portable because they can run off small, low-voltage batteries.

Not only radio receivers but a very large number of other electronic devices can be made much smaller now that transistors can be used. This has made a big difference to what is possible in space research.

Calculators, computers, electronic control mechanisms for industry and a wide range of instruments for communications depend on transistors. They are one of the most useful inventions of the century.

# MAKING THINGS BIGGER

We have said that a transistor works like any valve, controlling energies bigger than its own, just as a tap on a water main releases a great flood of water for the small work of turning it. Another way of looking at its action is to think of it as an amplification. The accelerator of a car amplifies the force of a man's foot to the force which accelerates both him and his car. A transistor amplifies a weak signal into one that can be detected. They do not, of course, multiply energy at zero cost. The extra energy must be fed in from petrol or from a battery. But the amplifier shapes the pattern of this extra energy, so that it is an exact copy of the small pattern that is fed into it.

*The electronic equipment in this satellite can be kept small because it makes use of transistors. They can run off the power supplied by solar cells which are also made of semi-conductors.*

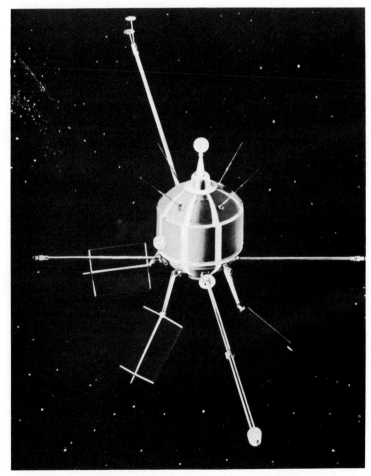

51

## FASTER AND FASTER

The years from 1951 to 1960 saw the development of another type of amplifier which opens up exciting possibilities, and again we find a crystal at the bottom of it all, this time a synthetic ruby crystal.

A transistor will control and amplify electric currents that vary not at all, or slowly, or at a frequency as high as a hundred million times a second. That seems an enormously high frequency, but the frequency of light waves by which our eyes function is more than a hundred million million vibrations a second. It is these frequencies which the new invention will amplify. It is called a LASER (lay-zer). The letters for the word stand for: Light Amplification by Stimulated Emission of Radiation.

If you would like to know more about transistors and lasers you will find some suggestions for reading at the

*A laser beam cuts the steel plate - and then? (James Bond in 'Goldfinger')*

end of this booklet. Here we will say very little. The ruby crystal of a laser is grown in the form of a rod, with two flat ends. One end is silvered to make it a mirror. The other end is partly silvered, so that it reflects some but not all of the light that falls on it.

The rod is surrounded by a spiral flash lamp that 'pumps' in the extra energy. The atoms in the ruby crystal absorb some of this energy, and give it out again at their own special frequency. The light given out by the ruby atoms is reflected to and fro between the end mirrors, disturbing other atoms, which then also give out light. The 'sideways' light escapes, so does some through the half-silvered end. But most of the light rushes to and fro in the crystal, and gets more and more intense until the part coming out in short bursts from the half-silvered end is a very powerful, narrow beam, all of the one special frequency.

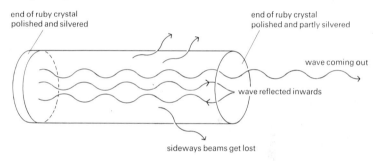

end of ruby crystal
polished and silvered

end of ruby crystal
polished and partly silvered

wave coming out

wave reflected inwards

sideways beams get lost

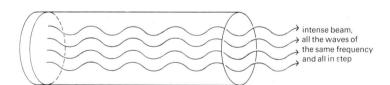

intense beam,
all the waves of
the same frequency
and all in step

*How a ruby laser works.*

High-frequency light waves affecting our eye produce the effect which we call violet; lower frequencies cause us to see red and the other colours of the rainbow come in between. A laser gives us a very intense beam, a very narrow beam and one that is accurately and purely of one colour. This is a tool which has enormous possibilities besides that of cutting James Bond in two!

## THINGS A LASER CAN DO

Some lasers work continuously with a power of only 4 watts. (The electric-light bulb in your sitting room is probably sending out light energy at the rate of 100 watts.) But crystal lasers that work in spurts can produce fantastic pulses of energy: the power of a single pulse may be 100 million watts and it can be concentrated upon a very small area, but it only lasts for a small fraction of a second.

Lasers have already been used for eye surgery and for cutting and welding metals with very high precision. Chemists can use them to trigger off changes in substances and so to investigate their structure. Because laser light is all one colour it can be used to measure very accurately how fast light travels. A laser beam has been bounced off the moon and back to earth.

*A surgeon operating on the retina of the eye with a laser.*

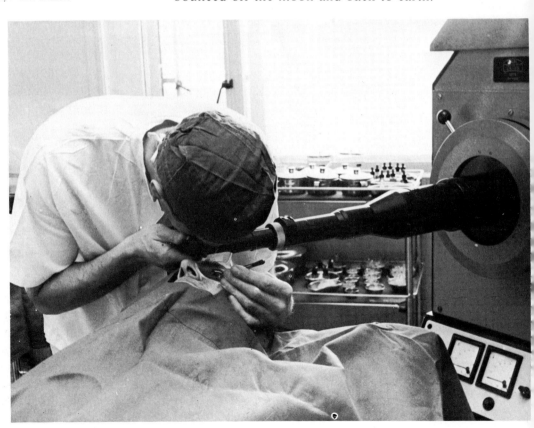

54

But the biggest possibilities probably lie in communications. Just as radio waves carry messages round the world so can laser beams. Because their frequency is higher it means they have a finer 'structure' and could carry many more messages at a time. There is a great deal of work still to do and many problems to solve, but they are certain to make changes in your life before many years.

### Something to work out

A laser beam takes 2.14 seconds to travel to the moon and back. Light travels at 300 million metres per second. How far away is the moon?

## LOOKING BACK AND LOOKING FORWARD

You will remember that we began by looking at some beautiful and curious natural crystals, wondering how they came to grow as they did and finding we could grow crystals ourselves under certain conditions. Then we looked at a wider range of crystals and saw how they could be classified according to their shape. A study of some of their properties made it possible to guess the inside story: the structure of crystals that makes them behave as they do. In the section 'Crystals for Today and Tomorrow' we looked at some of the useful jobs that crystals perform and how our guesswork about their structure gave us the power to make use of them.

This, if you stop to think of it, is the way in which science grows. Curiosity about the things around us is always the starting point. Afterwards we try to discover pattern and regularity, to classify what we have found out, and to write out 'laws' which are just general statements about the way things behave. But that does not take us very far.

It is only when we have the courageous imagination to guess at what lies behind the laws, and to invent a model that 'explains' what we see, that we are beginning to understand and to have some control. We use our model to experiment, to see if it goes on being useful and making sense of things we observe and the results we get. If it does, we go on using it and can turn our knowledge to practical account.

55

# A GRAIN OF MUSTARD SEED

The biggest changes in science often come from something that looks quite useless to begin with, troublesome perhaps, something small but curious, that leads one to wonder and guess; to test ideas, to reject them if they do not square with experiment and to begin again and again. Then, too, the worthwhile developments come almost always from team-work. Three men, two of them working with pencil and paper, and one at a laboratory bench, were chiefly responsible for the transistor. Lasers were developed by a team of three following up work that had been done by Charles M. Townes. Townes' work was based on the ideas of Einstein, Bohr and many others.

*Ali Javan, William R. Bennett Jr and Donald R. Herriott with the helium-neon laser, which was invented later than the ruby laser.*

We cannot see, when we begin, how useful it might be to come to a deeper understanding of, say, a tiny crystal, but two things are sure.

First, we shall never discover anything really new and valuable if we stop wondering about the things around us, if we cease to be driven by an insatiable curiosity.

Secondly, valuable discoveries will be lost if they are left in the research laboratories and never get passed on from the scientist to the engineer who has a flair for grasping at the same time both the fundamental ideas and their practical possibilities. Both have fascinating jobs that involve much hard work, perseverance in the face of difficulty and disappointment, as well as the thrill of success. Both must learn to co-operate with their fellow workers, to share ideas and problems, to encourage each other and rejoice at another's success. Both need the gift of a powerful imagination that is not afraid to leap into the dark. Would you like to be a scientist or an engineer?

*Ruby laser in action.*

# APPENDIX 1

## THE SYMMETRY OF A CUBE

We can cut a cube into two equal and opposite halves.
Like this          or like this          or like this

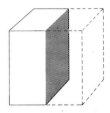 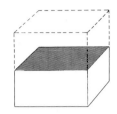 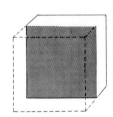

The shaded cuts are called planes of symmetry. So we have three planes of symmetry in a cube.

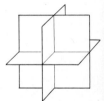

But we can also do it diagonally.
Like this          or like this          or like this

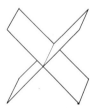 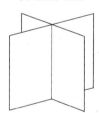

Here we have six more planes of symmetry, so a cube has nine planes of symmetry altogether.

If we put a stick through the centre of a cube like this we can turn the cube into four positions in which it looks exactly the same. Make a plasticine cube and try it. So we call this stick' a four-fold axis of symmetry.

We could put a stick in across the corners like this:

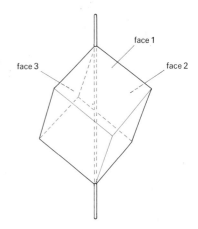

Can you see that you can turn the cube into three positions in which it would look exactly the same? This is a three-fold axis of symmetry. And here is a two-fold axis. Try it and see.

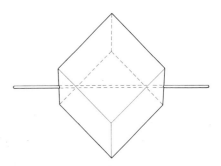

So a cube has thirteen axes of symmetry as well as nine planes of symmetry. There are three four-fold axes, four three-fold axes and six two-fold axes. When you have made up the models on page 21 see if you can find their planes and axes of symmetry. None of them has as many as a cube.

# APPENDIX 2

## THE FOURTEEN CRYSTAL LATTICES

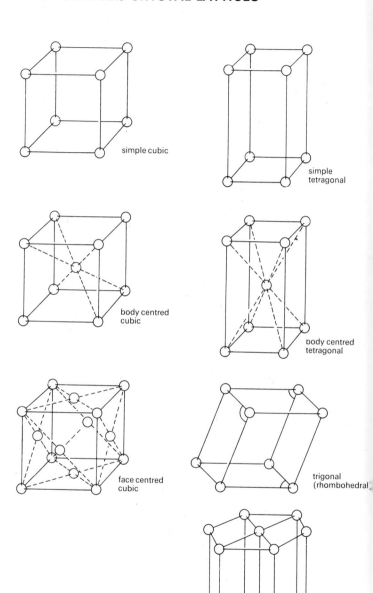

simple cubic

simple tetragonal

body centred cubic

body centred tetragonal

face centred cubic

trigonal (rhombohedral)

hexagonal

angles marked are less than right angles

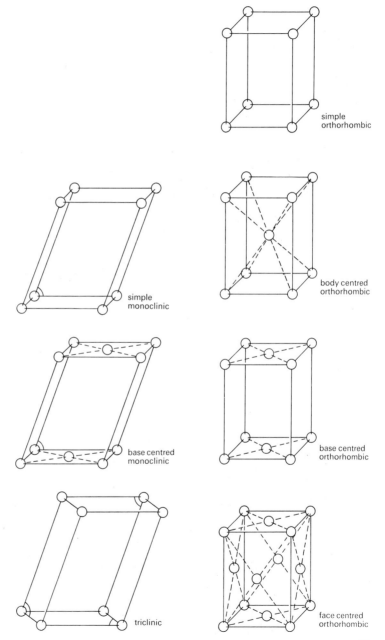

simple
orthorhombic

simple
monoclinic

body centred
orthorhombic

base centred
monoclinic

base centred
orthorhombic

triclinic

face centred
orthorhombic

61

# APPENDIX 3

## SOME PROPERTIES OF CRYSTALS AND COMMON SUBSTANCES

| Substance | Remarks | Page reference | Relative density | Hardness Mohs' scale | Colour | Streak† |
|---|---|---|---|---|---|---|
| Diamond | you might prove something was not diamond! | 13 16 36 | 3.5 | 10 | colourless yellow black | |
| Graphite | pencil lead; greasy feel | 24, 37 | 2.3 | 1–1.5 | black | black |
| Corundum | emery* paper | 18 | 4.0 | 9 | commonly grey or brown | |
| Magnetite | magnetic oxide of iron | 30 | 5.2 | 6 | black | black |
| Quartz | acid has no effect on it | 5 22, 24 44 | 2–6 | 7 | colourless smokey or tinted | |
| Calcite | three cleavages in different directions; fizzes under a spot of acid | 5 8 25 32 | 2.7 | 3 | colourless white brown yellow | |
| Biotite | in granite | 6 | 3.0 | 3 | dark brown | |
| Pyrite | fool's gold; in coal | 23, 24 | 5.0 | 6 | brassy | black |
| Haematite | an iron oxide, possibly earthy | -- | 4.8–5.3 | varies 1–6 | black | red-brown |
| Galena | lead sulphide; good cleavage in three directions | 30 | 7.6 | 2.5 | black | grey-black |
| Iron | tinned fruit cans are made of iron with a thin coating of tin | -- | about 7.5 | about 5 | grey | |
| Rock salt (halite) | try table salt | 7 40 | about 2.2 | 2.5 | colourless | |
| Sugar | several kinds | 42 | about 2 | 2 | colourless brown | |
| Hardwood | e.g. oak | | 0.6 | | light brown | varies |

\* Emery is a mixture of corundum and magnetite. Soak the paper, grind it and then the magnetite can be removed with a magn
† Streak is only a useful guide for metallic ores; most others give a pale or colourless streak.

# APPENDIX 4

## BOOKS FOR FURTHER READING

A. HOLDEN and P. SINGER, *Crystals and Crystal Growing*. Heinemann, 1961.
> This book will tell you in much more detail how to grow crystals of many kinds as well as much more about their symmetry and structure.

M. BYRNE, 'Growing Crystals', *School Science Review*, March 1967.
> Your teacher almost certainly has a copy of this article.

K. W. MARRISON, *Crystals, Diamond and Transistors*. Pelican, 1966.
> The author says that this book 'should enable sons and daughters to put more unanswerable questions to their parents than before', so you might like to try it!

MARTIN MANN, *Revolution in Electricity*. John Murray, 1962.
> A fascinating book, especially on the new technology arising from solid-state discoveries.

D. W. HARDING and L. GRIFFITHS. *Materials*. Longmans.
> Another book in this series that will tell you about other solids, some noticeably crystalline and some not.

J. and A. BAUM, *A Book of Jewels*. Paul Hamlyn, 1966.
> Borrow this book from the library and enjoy the superb colour photographs.

NUFFIELD FOUNDATION, *Making Diamonds* and *Growing Crystals* (Chemistry background books). Longman–Penguin, 1968.